From Human Dawn
to Politics

by Rolf A. F. Witzsche

Contents

About the Illustrated Science series 7

*The dawn of humanity, civilization, and God 10

Only during Ice Age environments 11

The development of the human species didn't even begin 12

The critical factor that dominates the timing 13

A type of nourishment for mental development 14

Cosmic-ray particles 15

The interaction between cosmic-ray flux and human development 17

Since it became possible with Carbon-14 measurements 18

Great progressive developments in civilization 19

The Maunder Minimum 20

In Palaeozoic history 21

Increases in cosmic-ray flux, ...23

The rise and fall of spiritual recognition 24

The warm periods stand out as periods of cultural destruction 25

With the higher-resolution Carbon-14 measurements 26

Beyond the insanity of the Nazi holocaust 27

The Sleep of Reason Produces Monsters 28

The future of humanity is not inherently bound to the cosmic default 29

Humanity has in its path become a highly developed species 30

This means that no huge feat is required for humanity to free itself 31

The biofuels holocaust can be stopped in a similar manner 32

Only the Ice Age Challenge cannot be so easily met 33

The critical breakthrough can be made in a short time nevertheless ...34

*The truth is dead 35

But will she rise again? 36

This is the truth 37

*The Giza pyramids and Stone Henge 38

Grasping the fire 39

Some of the wonders stand as giant structures 40

In asking these questions 41

The builders of the pyramids 42

Built 12,800 years ago, and not as tombs 43

The complete absence of inscriptions 44

Built around astrophysical phenomena 45

*Images of the Primer Fields in action 46

The type of image that we see today in the Red Square nebula 47

The famous three stars 48

Three stars determined the pyramids 49

Three pyramids as a single project 50

*A silent testimony 51

Another significant story.. 52

Giza is located near Cairo... 53

The Sahara was a lush region.. 54

Northern boundary of the ice age safe zone 55

*The safe line 56

A perfect zone free of hurricanes 57

Floating agriculture is small stuff... 58

The future of Canada, Russia, and Europe...................................... 59

*The Stone Henge project 60

A ring of 56 chalk pits.. 61

A lab experiment, published in 2003 62

A ring of 56 distinct plasma filaments... 63

A very-high power plasma stream 64

It is surprising to note 65

We no longer see these patterns 66

*Politics versus science 67

Humanity urgently requires the ice age challenge 68

A spiritual task, more than a political task 69

We wield weapons out of weakness 70

We face a choice 71

Humanity still lives in the Roman age 72

We simply starve them to death 73

*The ice age challenge as a new paradigm 74

Only we lose by our failing 75

To rekindle that flame in the heart 76

Not even the sky will be a limit 77

Will we dare to step up onto the wings and fly? 78

*The greatest danger that we face 79

Brainwashed by the choruses of the professional scoundrels 80

People who have sold their soul for a song 81

Brainwashing to keep the internal-fusion sun theory alive 82

Children to become depopulated? 83

The Sun is an electrically powered star....................................84

Behind the scene of denial..85

The intelligence of humanity 86

Break through the fog of political games..................................87

Alert scientists and truth-seekers 88

Scientific progress the hallmark of humanity 89

Unlimited energy resources 90

*Potentials stand before us right now 91

Maybe this is what the Universe recognized and put to our credit 92

To create a society without empire 93

5

The city on a hill 94

The greatest economic and scientific development 95

Warm climates after the Little Ice Age 96

The war against empire has not yet been won.................................... 97

Society's self-directed spiritual development...................................... 98

To meet the human need 99

 *Challenge of the dimming Sun.. 100

Something to celebrate.. 101

Harvest is Seedtime 102

Discovering Love 103

The melody of nature - what a song!................................... 104

Lu Mountain.. 105

Listen to the song.. 106

Flight Without Limits... 107

Brighter than the Sun.. 108

About the Illustrated Science series
On the Ice Age and Climate Change
and the book

From Human Dawn to Politics

Book 4 of the series: Ice Age of the Dimmer Sun in 30 Years

Ice Ages have evidently played an important role in the development of humanity, since the last 2 million years of the history of humanity occurred during the modern Ice Age Epoch, the Pleistocene Epoch. For 85% of this time glaciation conditions occurred. What we refer to as 'history', spans only the brief period of the current interglacial. The period began with building of the Giza Pyramids 12,800 years ago, as some researchers say. The extremely accurate alignment of the pyramids confirms that potential. At the break-out from the last Ice Age, major features of the 'Primer Fields' that focus plasma onto the Sun, would likely have been visible in the sky at this time and be used for celestial orientation. Likewise, the Stonehenge monument in England reflects features that are visible in high-energy plasma discharge experiments; which may have been visible in the sky in early times in the post glacial period.

The great monuments suggest that the ancient builders were highly intelligent, which may reflect conditions in Ice Age environments. During the modern interglacial, the great cultural developments occurred during the cold 'little' ice age periods, which are periods of high rates of solar cosmic-ray flux. The 'insane' periods of modern politics where periods of opposite conditions.

During the inactive state in solar activity, when the Sun reverts to a type of cosmic default level with 70% less radiated energy, higher rates of solar cosmic-ray flux are being experienced, with a reduced shielding effect by the 'thinner' plasma surrounding the Sun. We are presently on track back to those conditions. At the present rate of diminishment, the solar activity phase-shift threshold to the next Ice Age period may be crossed in 30 years, or in the 2050s, most likely. With the primer system gone inactive, the climate on Earth will get 40 times colder than the Little Ice

Age in the 1600s had been. Ice core evidence promises that. Without the needed preparation for human living in such an environment, 99% of humanity would die of starvation, both by the cold, and by CO_2 depletion that diminishes agriculture, as more CO_2 becomes dissolved into the sea.

With the 'Primer Fields' being critical for our very existence, the exploration of them is likewise critical.

In the Little Ice Age, between 10% and up to 30% of the populations in Europe had perished by starvation. The last Big Ice Age was evidently vastly harsher. Only 1-10 million people emerged from it alive. That's all we had after 2 million years of development. We want to do far better this time around; and we can, with large-scale technological infrastructures for our food supply. But will we create them? Will we get the job done in the 30 years that we still have left before the Ice Age starts anew? Will we even consider it? And how certain are we that the phase shift to the next glaciation period will begin, as the evidence suggests, in the 2050s? We have no slack on this front. Should we fail us on this absolute front, we would be committing suicide.

Numerous fields of evidence tell us that the next Ice Age is near. That's where the truth begins. Most of the evidence was discovered in the 1990s and thereafter. Some evidence is measured in ice cores; some is measured in space, by satellites. Some measurements are also made on the ground in terms of measurements of the Earth's magnetic-pole drift observed in northern Canada. All of this is seen combined with high-energy physics experiments at a leading national laboratory, and is also explored in the small in static experiments.

So, what will the answer be? Will we move with the evidence? Or will we lay ourselves down to die by default?

It takes an independent researcher to brake the taboos that have kept mainstream cosmology imprisoned, increasingly, during the past century, even while what is regarded as taboo is known to be wrong.

The Illustrated Science series is intended to open the scene beyond the threshold of accepted taboos, to where the actual physical evidence speaks for itself.

The scope of the existential challenge that the Ice Age brings with it, takes astrophysics out of the academic domain and places it into the foreground as one of the most-critical issues of our time. The big Climate Change events that have already worldwide effects are mere fringe effects in the flow of the ever-changing cosmic dynamics. The big effect, when the Ice Age begins anew, promises to be caused by a dimmer and colder Sun. The loss of 70% of the Sun's radiated energy defines our climate future that begins in the near term.

Sure, we can live with all that by creating new platforms for agriculture that are able to operate under Ice Age conditions. But will we do it? The task is enormous. Or will we fail ourselves on this front? We have no reason to allow us to fail. We have the materials and energy resources on hand to accomplish everything that is required for us to continue to live in an Ice Age World. But will we do it? The big question that never goes away, therefore, is; will we develop our inner resources as human beings sufficiently to get the job done, and to get it done in time? Or will we do nothing, ignore the challenge, and condemn our children and one-another to an agonizing death by starvation? That's the choice.

Towards meeting the inner challenge, I have created the epic series of novels, The Lodging for the Rose. And further, towards meeting the science challenge, I have produced numerous research books and several dozen exploration videos that the Illustrated Science series is modeled after. The work is the result of a quarter century of research, for which numerous elements of evidence in related fields came to light during the timeframe of my research.

It is my hope that the work that went into all of these projects will help in some degree - for humanity that we are all a part of - to write itself a ticket to have a future.

High-resolution color images, of the images in this book, can be obtained at www.iceagetheatre.ca

The dawn of humanity, civilization, and God

Only during Ice Age environments

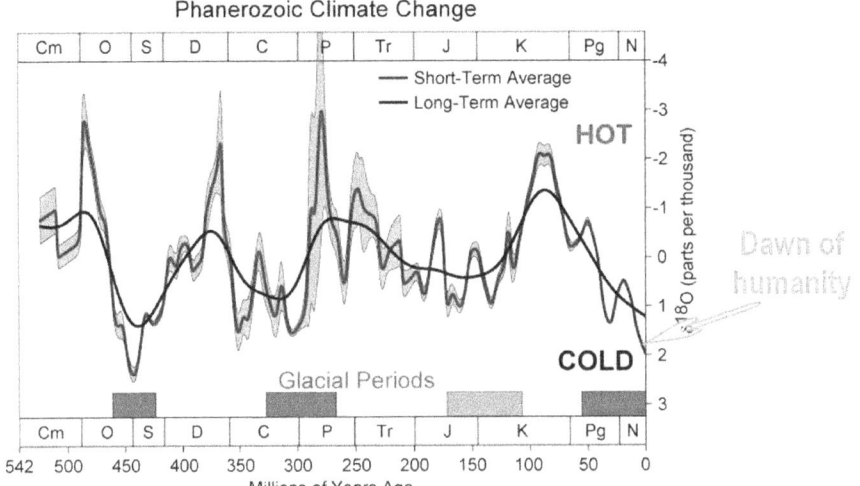

Our development into a highly advanced and capable species with creative and productive powers that no other form of life can equal, may have been made possible by one of the special conditions that exist on the Earth only during Ice Age environments.

The development of the human species didn't even begin

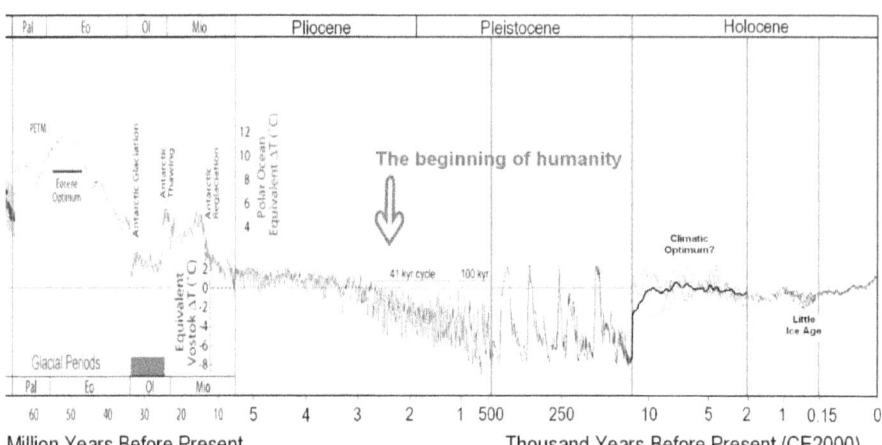

The development of the human species didn't even begin until the modern ice age epoch began roughly two million years ago.

The critical factor that dominates the timing

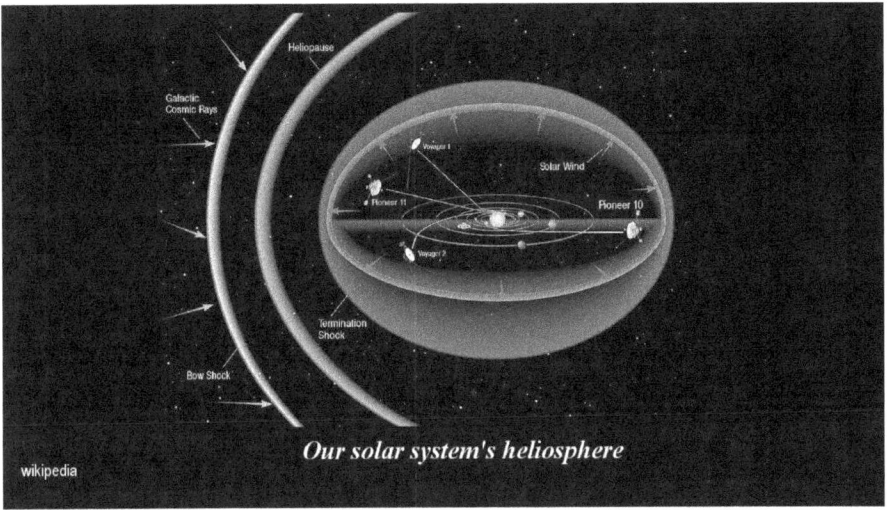

Our solar system's heliosphere

The critical factor that dominates the timing of the development of humanity, is evidently the increased galactic cosmic radiation that results during glaciation conditions when the shielding effect of the heliosphere vanishes. When the Primer Fields collapse, the spherical shell of plasma that the solar winds create around the solar system, will vanish in short order. The plasma shell is called the heliosphere. It forms when the solar winds grind to a halt, far from the Sun, whereby the plasma of the solar winds accumulates from within, into a relatively dense shell.

Without the heliospheric plasma shell surrounding the solar system, the full force and volume of the galactic cosmic-ray flux will then be able to penetrate to the Earth.

The penetration of the cosmic-ray flux, and its interaction with our biological systems. appears to have a beneficial effect.

A type of nourishment for mental development

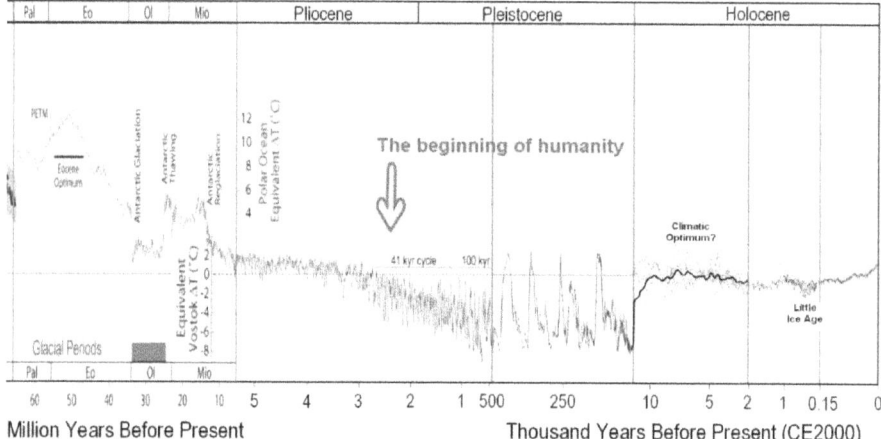

Temperature of Planet Earth

Source: Wikipedia

The beginning of humanity

This means that the universe has given us a great gift by providing two distinct environments that in conjunction have aided the development of humanity on two critical fronts alternately. The glaciation periods furnish an environment that provides a type of nourishment for mental development, with a high rate of galactic cosmic-ray flux, while the interglacial warm periods furnish an environment for easy living with plenty of food available, under a brilliant Sun. Both aspects appear to be needed for the advance of human development. It may be that the dawn of humanity did not begin until both conditions were established.

The critical conditions simply didn't exist until the ice ages began.

14

Cosmic-ray particles

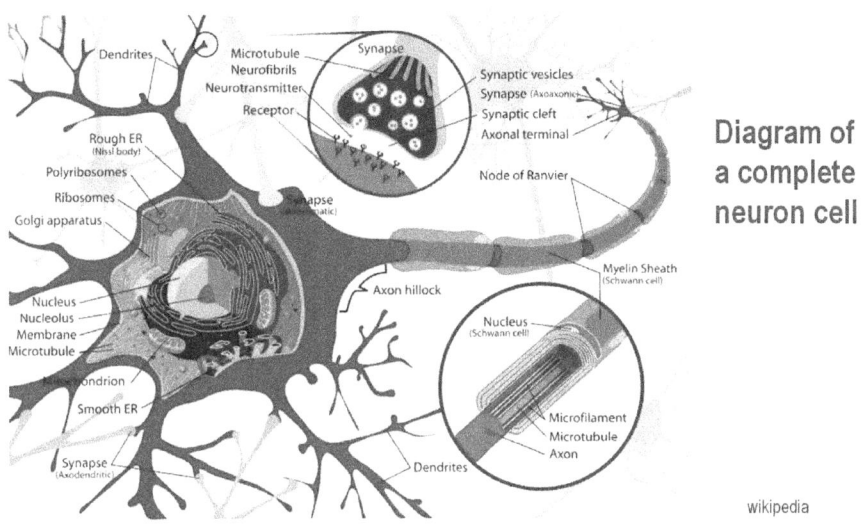

Diagram of a complete neuron cell

wikipedia

As I said before 'cosmic rays' are not 'rays' in the standard sense, like rays of light, but are high-energy electrons and protons in motion at upwards to the speed of light. With the cosmic-ray particles being electrically charged, they typically pass right through our biological systems without damaging anything. However, as they path through the biological system, they generate electric currents by induction that appear to be beneficial for the high-level neurological functioning that governs the complex human biology. They tend to facilitate critical functions that apparently might not occur otherwise.

Our biological and neurological systems, are extremely complex electric systems, which evidently benefit from increased electric activity, especially so when it is extended over long periods of time. Our cognitive, scientific, and what may be called, spiritual powers, appear to be derived from this high-level-type of electric nourishment. The cosmic-ray interaction may have enabled cognitive powers, and then aided the continuing development of them, by which we became a high-level species.

The interaction between cosmic-ray flux and human development

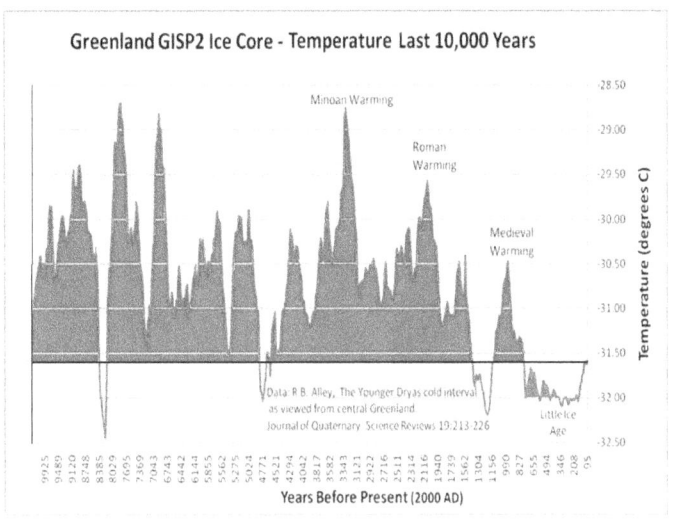

The interaction between cosmic-ray flux and human development is surprisingly evident when we look back into the history of civilization as it developed in the current interglacial period where some cosmic-ray flux is received from the Sun, primarily during the cold periods.

Since it became possible with Carbon-14 measurements

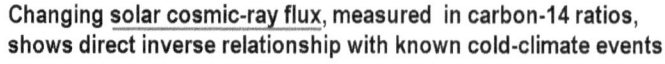
Changing solar cosmic-ray flux, measured in carbon-14 ratios, shows direct inverse relationship with known cold-climate events

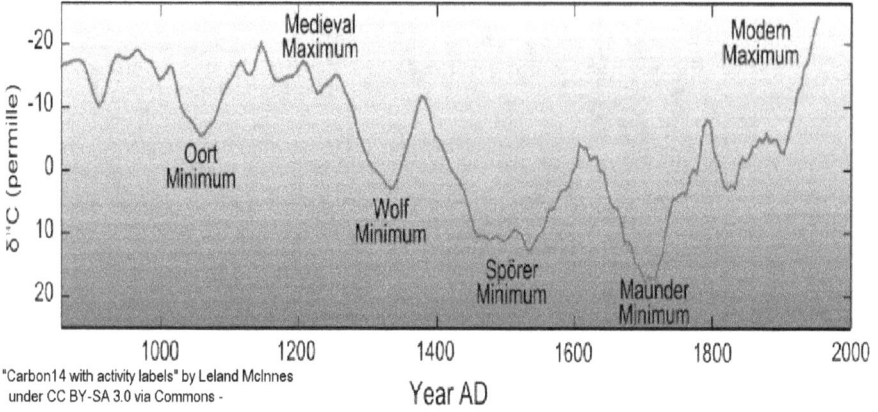

"Carbon14 with activity labels" by Leland McInnes
under CC BY-SA 3.0 via Commons -

Since it became possible with Carbon-14 measurements, to measure the historic volumes of the solar cosmic-ray flux, and it became self-evident that this volume varies dramatically with changing solar activity, which is also mirrored in temperature changes, it becomes possible to correlate cultural effects with changes in solar cosmic-ray flux.

Great progressive developments in civilization

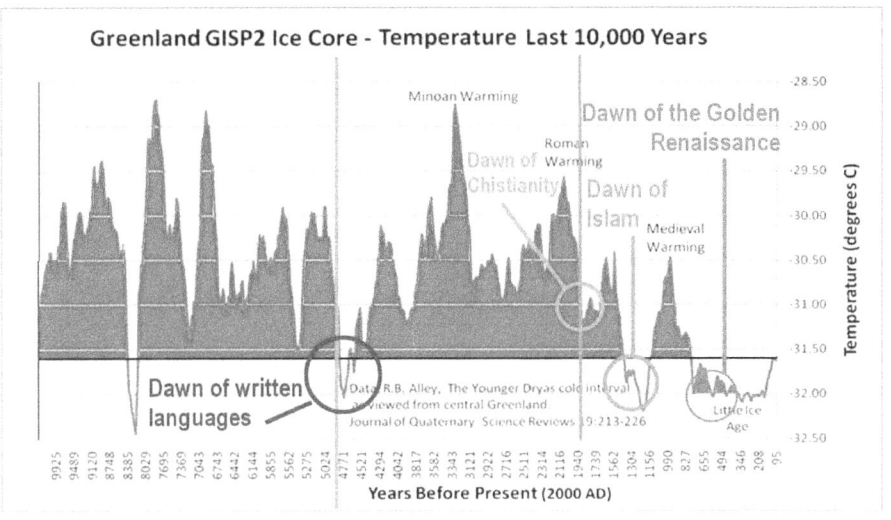

As we make the comparisons, it becomes surprisingly evident that the great progressive developments in civilization all occurred during the cold periods, which are periods of high volumes of solar cosmic-ray flux.

The Maunder Minimum

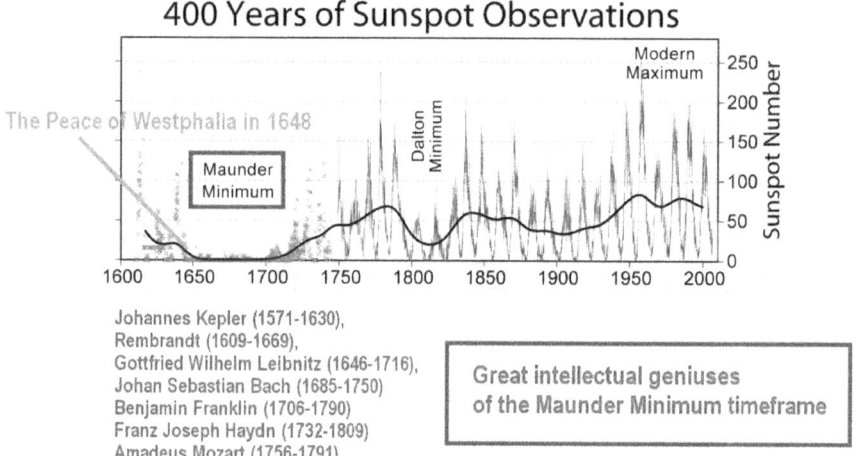

400 Years of Sunspot Observations

Johannes Kepler (1571-1630),
Rembrandt (1609-1669),
Gottfried Wilhelm Leibnitz (1646-1716),
Johan Sebastian Bach (1685-1750)
Benjamin Franklin (1706-1790)
Franz Joseph Haydn (1732-1809)
Amadeus Mozart (1756-1791)

Great intellectual geniuses
of the Maunder Minimum timeframe

For the same reason was the period of the Maunder Minimum of low solar activity - where no sunspots have occurred - which is coincident with the Little Ice Age - a period in which enormous scientific and cultural development has occurred. During the Maunder Minimum, the extremely high solar cosmic-ray flux of that period gave us the greatest peace treaty of all times, the Treaty of Westphalia that still stands as the foundation for civilization. The period also gave us the great musical geniuses and scientific geniuses that are still admired today.

In Palaeozoic history

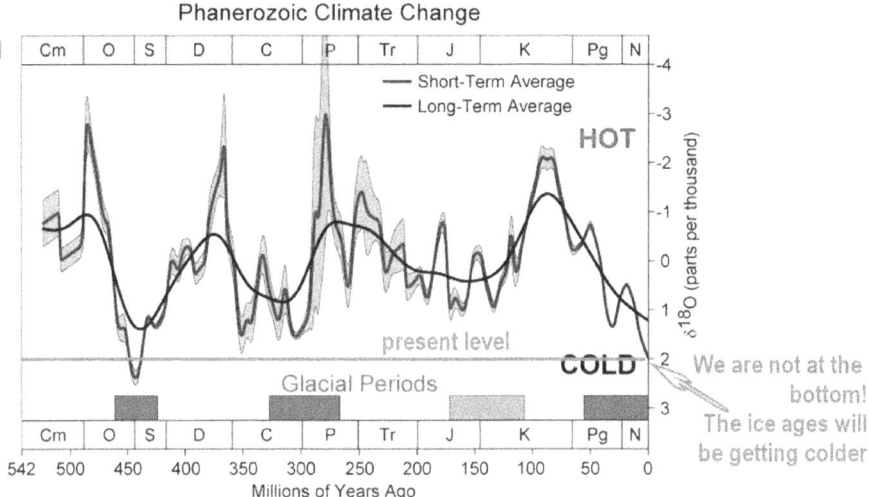

Phanerozoic Climate Change

As I said before, the conditions in Palaeozoic history, when the Primer Fields for the Sun collapse and the Sun turns off, are rare. They occurred only four times, as we have records of them, with the earliest having occurred roughly 450 million years ago.

There my have been earlier, deeper glaciation periods between 650 and 750 million years ago, in which the entire Earth froze up into a giant snowball and remained frozen for tens of millions years of until the Sun became reactivated again. While the snowball-earth theory is controversial, it is well within the range of normal possibility in the context of the Primer Fields dynamics.

Rare as the great glacial periods may be, our existence is linked to them. The coincidence of the dawn of man with the modern Pleistocene ice epoch is significant, as it indicates that our very existence as a highly developed species may be the direct result of the potentially very high cosmic-ray flux density that occurs in times when the solar heliosphere does not exist.

Since the normal 'rich' conditioning for human development gets interrupted by the interglacial periods, which is all that we have known, we really don't know then what 'normal' living is like, even as we are about to become drawn into it again.

Increases in cosmic-ray flux,

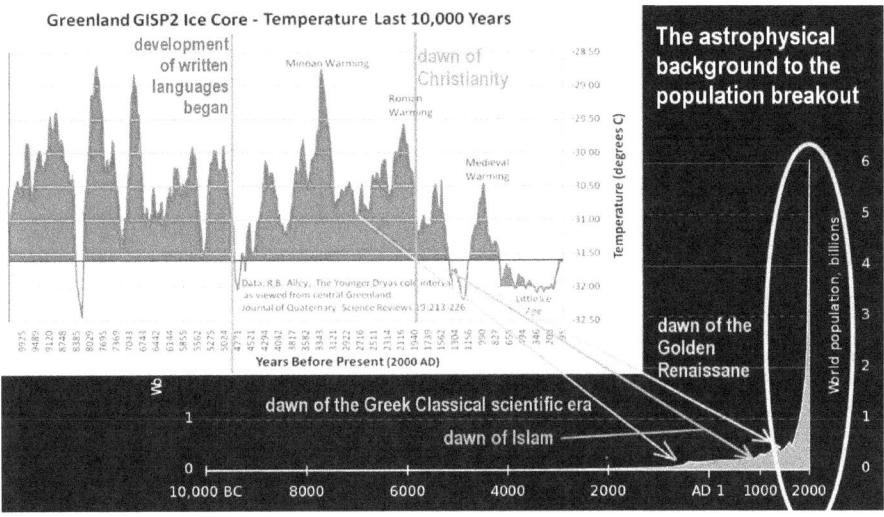

During the lean period of the interglacial, we have seen only occasional increases in cosmic-ray flux, this time coming from the Sun. History tells us that in the few occasions of high volumes of solar cosmic-ray flux, amazing cultural developments occurred. Almost all of the great cultural breakthroughs were made in these types of times. The development of written languages, for example, occurred in one of the deep cold times with high volumes of solar cosmic-ray flux.

History also tells us that the great developments that did occur in these cosmic-ray rich times, were typically cognitive and scientific in nature, which renders them to be forms of spiritual development that may be termed the pinnacle in the mental realm.

In a very real and powerful way, the rise and fall of civilization follows the ebb and flow of the recognition of spiritual values in society.

The rise and fall of spiritual recognition

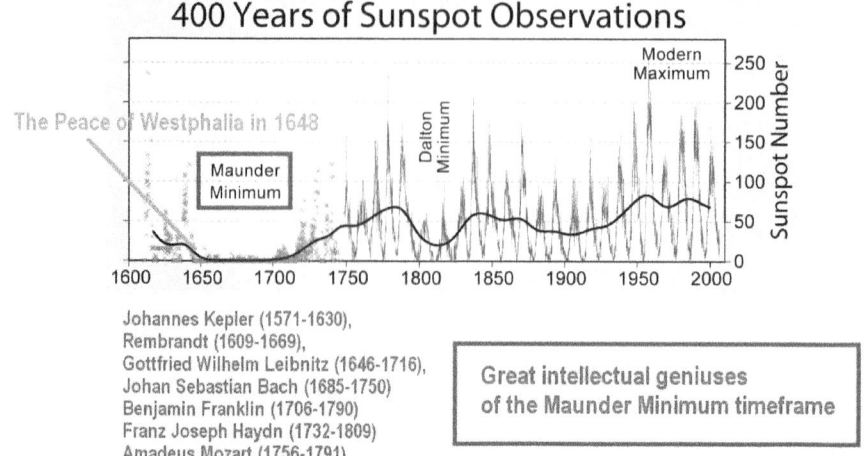

400 Years of Sunspot Observations

Johannes Kepler (1571-1630),
Rembrandt (1609-1669),
Gottfried Wilhelm Leibnitz (1646-1716),
Johan Sebastian Bach (1685-1750)
Benjamin Franklin (1706-1790)
Franz Joseph Haydn (1732-1809)
Amadeus Mozart (1756-1791)

Great intellectual geniuses
of the Maunder Minimum timeframe

The rise and fall of spiritual recognition in turn, follows the historic rise and fall of the cosmic-ray flux reaching the Earth, which is typically the inverse trend of solar activity.

The warm periods stand out as periods of cultural destruction

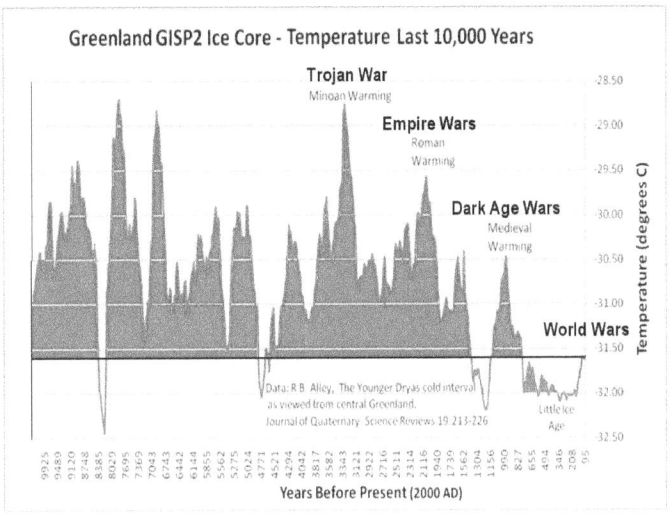

The inverse also proves the principle further. Just as the cold periods were periods of great cultural achievements, the warm periods stand out in the long sweep of history as periods of cultural destruction, periods of war, from the Trojan War, to the Roman Wars, to the Colonial Wars, to the World Wars and so on.

With the higher-resolution Carbon-14 measurements

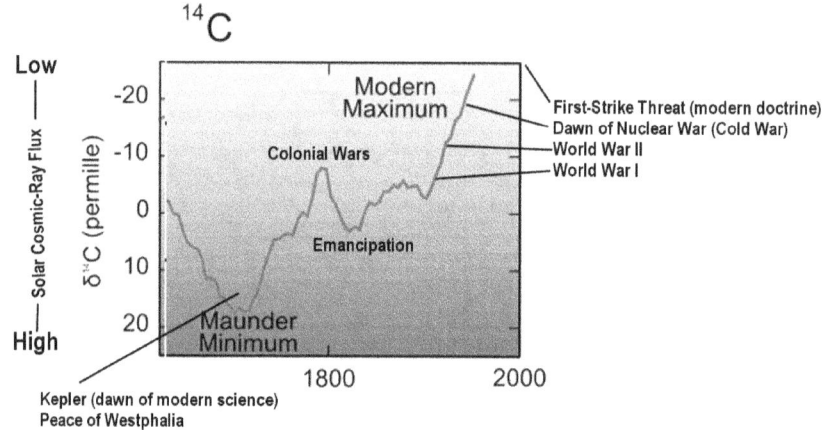

The same correlation is also evident in more recent times, in the context with the higher-resolution Carbon-14 measurements of solar cosmic-ray flux. The diminishing solar cosmic-ray flux from the time of the Maunder Minimum to the strong solar time near the year 2000, has been a period of accelerating collapse in civilization, a period of increasing insanity that became reflected in world wars, nuclear-war terror, and the modern hair-trigger stand-off towards the now fully prepared for, first-strike thermonuclear destruction of all life on Earth, for which the decision time on warning has been shrunk to roughly one minute, and the execution time to roughly one hour.

That's the present state. If this isn't utter insanity, what would qualify for the term?

Beyond the insanity of the Nazi holocaust

We have drifted far beyond the insanity of the Nazi holocaust that had murdered 6 million people in six years of fascist madness. Modern society exceeds this 100-fold. It is murdering upwards to 100 million people per year in the biofuels holocaust, quietly, unseen, with the sword of starvation forged by the mass-burning of food in a starving world. The amount of diverted agricultural resources that goes into the project to be burned, would normally nourish 400 million people. This adds up to supreme genocide in a world that has a billion people living in chronic starvation.

The Sleep of Reason Produces Monsters

The Spanish painter Francisco Goya illustrated the connection in his etching 'The Sleep of Reason Produces Monsters', of the Caprichos series of 1798.

It appears, by what we see happening all around us, that the pinnacle of our humanity, our cognitive and spiritual capability, and also the lack thereof as the case may be, is deeply affected by the every-changing cosmic-ray flux that the Primer Fields stand in the background to as an element of the larger scene.

The future of humanity is not inherently bound to the cosmic default

Francisco Goya (1746–1828) -
Disasters of War series -
Wikipedia

However, the future of humanity is not inherently bound to the cosmic default for the mental environment that the cosmic-ray environment dishes up. The sleep of reason is not our inevitable fate. The sleep can be ended. The age of reason can be restored.

Humanity has in its path become a highly developed species

The School of Athens - fresco by Raffaello Sanzio (1511) at the Vatican - Wikipedia

Humanity has in its path become a highly developed species with far greater capabilities than society gives itself credit for.
Of course it takes some active commitment by society to raise itself above the 'sod,' and consciously become that human factor in the world that it has the potential to be. There is no need for a society of human beings to allow events to grind it down, when it has the power to generate its own events. Nor is there a need for humanity to wait for changing conditions to raise it up, when it has the potential to raise itself up, to create its own conditions, and to make this happen decisively.

This means that no huge feat is required for humanity to free itself

Annihilation is assured

500,000 times
Hiroshima
in one hour

Castle Bravo - the first U.S. test of a dry fuel thermonuclear hydrogen bomb - March 1, 1954 at Bikini Atoll, Marshall Islands

This means that no huge feat is required for humanity to free itself from the threat of thermonuclear annihilation, that is now fully prepared for. The feat to end this impending Armageddon can be accomplished in a week. It would take less than a day to remove the trigger-happy rulers from power who wield the club of war and terror, and own the button for Armageddon. After that is done, it wouldn't take more than a week to physically disable and dismantle the entire nuclear-war machine anywhere in the world. To accomplish this is not a huge physical task, and by taking these steps, humanity would write itself a ticket to have a future, which presently isn't even a concept anymore.

The biofuels holocaust can be stopped in a similar manner

Mass Murder with Biofuels
a YouTube video

El Tres de Mayo, by Francisco de Goya - Wikipedia

The biofuels holocaust can be stopped in a similar manner, and just as fast. All the doctrines that stand behind the holocaust are built on lies, from the manmade global warming doctrine to the depopulation doctrine. The mass murder in the world can be completely ended in a month and sanity be restored again. No gigantic feat is required for society to rebuilt its humanity that way. On this path, society would gain more than just a little self-respect.

Only the Ice Age Challenge cannot be so easily met

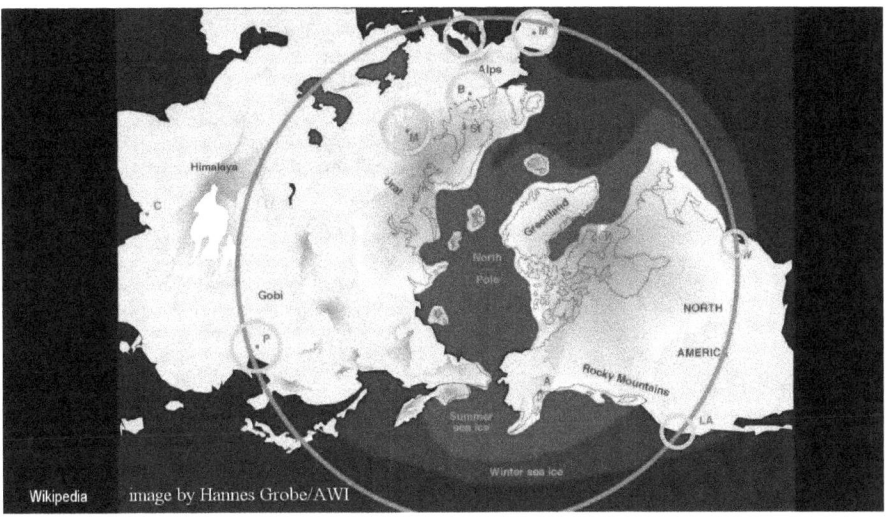

Only the Ice Age Challenge cannot be so easily met. It will take 30 years for society to build the infrastructures to relocate all countries outside the tropics, into the tropics, before their territories become uninhabitable.

The critical breakthrough can be made in a short time nevertheless

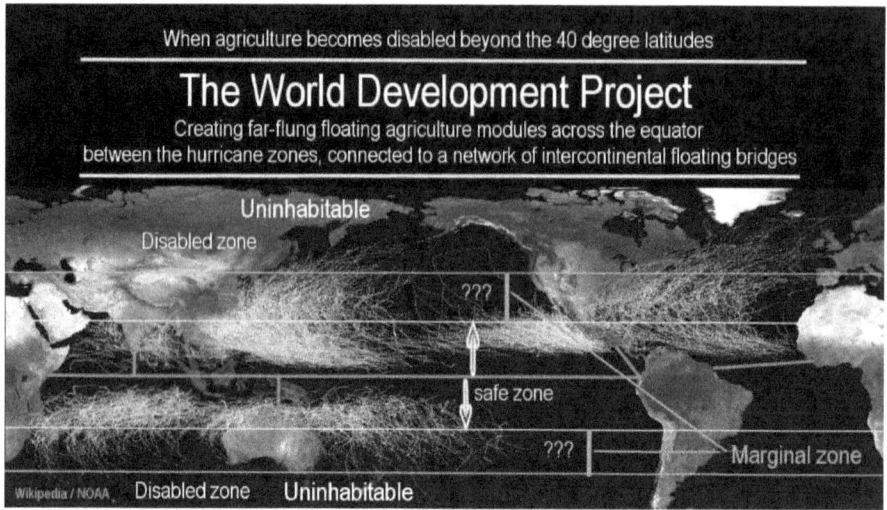

When agriculture becomes disabled beyond the 40 degree latitudes

The World Development Project

Creating far-flung floating agriculture modules across the equator
between the hurricane zones, connected to a network of intercontinental floating bridges

However, the critical breakthrough that is needed to get the ball rolling, can be made in a short time nevertheless. No miracles are needed. The evidence is plain. The imperative is hugely impelling. The power to meet the challenge exists in society. In taking up the challenge and moving with its imperative, humanity would give itself and its children a chance for certain survival. The concept, presently, appears to be almost non-existing.

*The truth is dead

Francisco Goya:
The Truth is Dead
from the series:
Disasters of War
(1810-1820)

As Goya had put it in his three-part etching, "The truth is dead"

But will she rise again?

Francisco Goya:
Will She Rise Again
from the series:
Disasters of War
(1810-1820)

"But will she rise again?" he asks.

This is the truth

Francisco Goya:
This is The Truth
from the series:
Disasters of War
(1810-1820)

"This is the truth," he says. She is the truth that defines us all, and this truth is alive. It is us, potentially. More than this we do not need.

The Giza pyramids and Stone Henge

Grasping the fire

The Greatest Miracle on Earth
a human being
born in the image of God
reaching for spiritual light
to understand the universe
and itself

The development of humanity appears to have always been intertwined with 'grasping the fire,' both mentally and physically. On this road humanity wrought many great wonders throughout the long history of its self-development.

Some of the wonders stand as giant structures

Some of the wonders stand as giant structures that still puzzle us today. How were they built? What where they built for?

In asking these questions

In asking these questions, some of the ancient structures that we marvel at, ironically, also come to light to serve as a link to the future. They give us a hint of what the ancient builders may have seen in the sky, which can no longer be seen, because the modern age unfolds in an electrically collapsing solar environment that is trending towards its impending solar cut-off point.

The builders of the pyramids

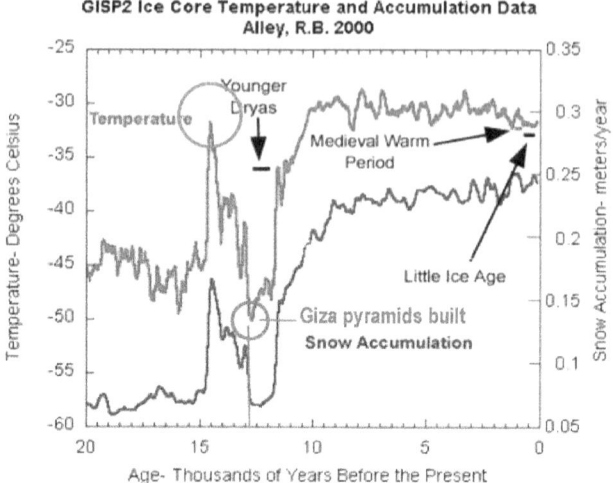

The builders of the pyramids may have experienced both of the solar extremes: A more brilliant Sun than we have today; and the inactive Sun that would have created the Younger Dryas period of renewed glaciation, in which researchers suggest the Giza pyramids have been built.

Built 12,800 years ago, and not as tombs

Contrary to assertions by Egyptology, the Giza pyramids were likely built 12,800 years ago, and not as tombs, for which they were later used. The distant date is derived by running the astrophysical clock backwards until a point is reached at which the Sphinx of the Giza complex is seeing its own image in the sky on a solstice sunrise.

The complete absence of inscriptions

Grand gallery of the Great Pyramid of Giza

The suggested date seems to be confirmed by the complete absence of inscriptions inside the pyramids. Inscriptions became customary much later, during the time of the pharaohs. Numerous other features likewise place the Giza pyramids far outside the timeframe of the pharaohs.

One of these features is the quality of their design, and the precision in construction. These place the Giza complex distinctly into a category of its own that nothing which the Egyptians had produced, and had been capable of at the time, comes even close to.

Built around astrophysical phenomena

Assuming that the pyramids were built 12,800 years before the present, which is the most reasonable estimate to date, what would have been the motive for the people at the time to build the giant structures that they built?
It appears that the motive may have been religious in a sense, while it was built around astrophysical phenomena.

The Red Square Nebula

During the great warming period, 2000 years previous to the building of the pyramids, a very large Dansgaard-Oeschger event broke the deep chill of the last ice age. At this time when the Sun became extremely active, the people may have seen with their naked eye the plasma images of the Primer Fields in action, centered on the Sun.

The type of image that we see today in the Red Square nebula

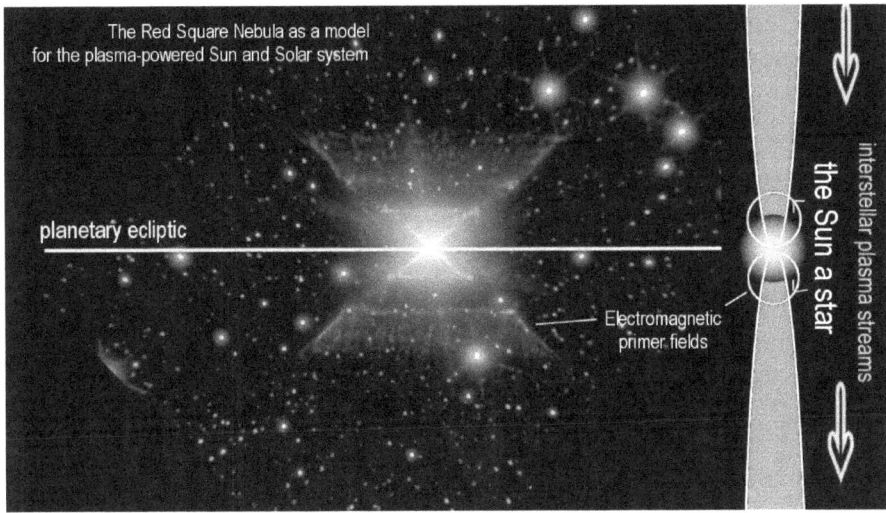

The type of image that we see today in the Red Square nebula may have been seen 'bright' and clear in the ancient skies. The full image may have been seen in the time of a solar eclipse, while a part of the image may have been 'blazing' in the night skies from dusk to dawn.

The people may have seen in the sky two pyramid shapes of light, with the Sun between then. On this basis the Sun may have became an object of worship the giver of abundant life resting on a pyramid. Then, suddenly, the Sun becomes inactive again. The idea must have emerged at this time to build a pyramid for the Sun - a house for the gods, provided on the Earth

When a thousand years later the Sun became brilliant once more and soon thereafter went inactive again, a deep crisis may have erupted that inspired the building of the greatest pyramid of all times to entice the Sun god to restore its shiny face for his people.

The famous three stars

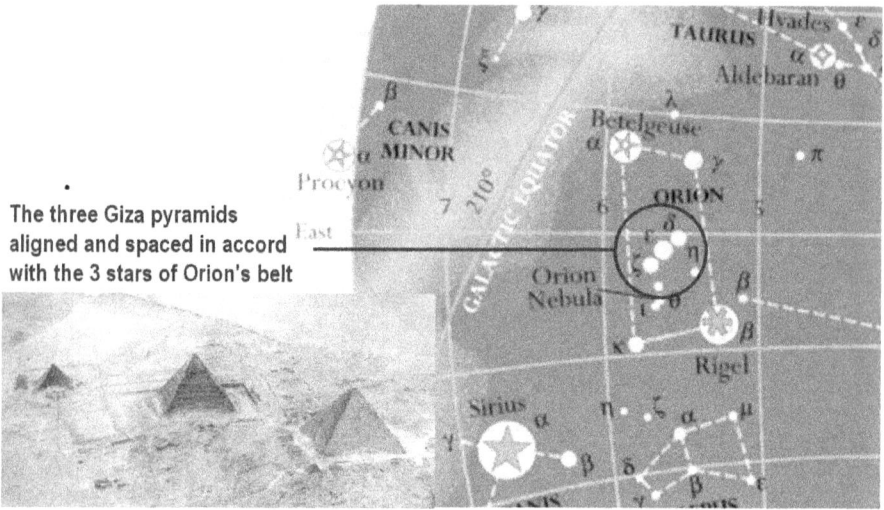

The three Giza pyramids aligned and spaced in accord with the 3 stars of Orion's belt

The people may also have seen among the plasma images in the sky, perhaps right at the center of them, the famous three stars of the belt of the constellation Orion.

Three stars determined the pyramids

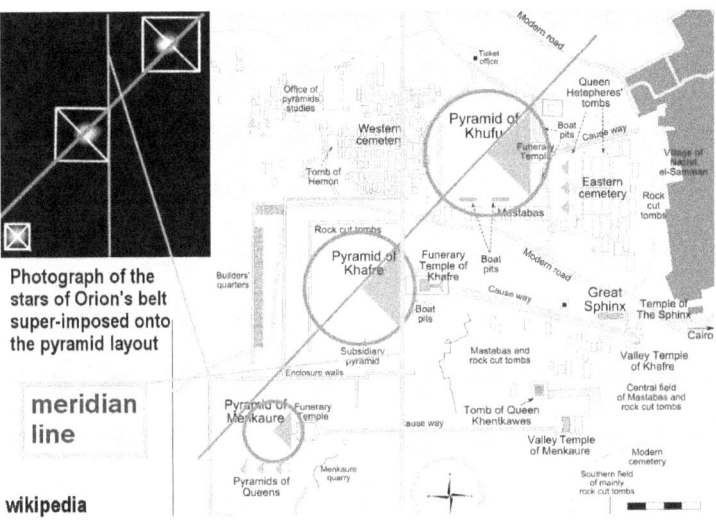

Photograph of the stars of Orion's belt super-imposed onto the pyramid layout

meridian line

wikipedia

These three stars appear to have determined the positions of the pyramids, their alignment to each other, and their relative size.

Three pyramids as a single project

When the great pyramid building project was launched, it appears that the ancients did not built not just one pyramid, to inspire the gods, but have built all three pyramids together as a single project, with each pyramid being exactly aligned and proportioned in accord with the three stars of the belt of Orion.

*A silent testimony

In this manner the pyramids of Giza stand for us as a silent testimony that the brilliance of the Sun is not a permanent feature, but is a fleeting phenomenon at times, which once had inspired a monumental response by a 'nation' to get the shiny face of the Sun back.

Another significant story

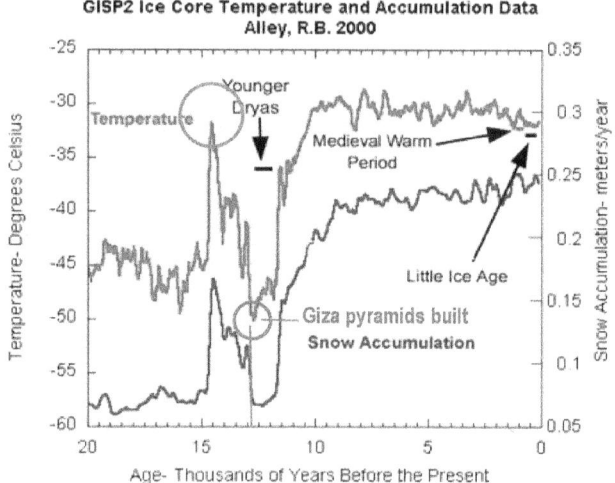

The pyramids also tell us another significant story. The evident fact that the Giza project was built 12,800 years ago is important, as it tells us that the region in which it was built was biologically sufficiently strong to support a large population.

Giza is located near Cairo

Giza is located near Cairo on the thirty degree latitude in the northern part of the Sahara.

The Sahara was a lush region

 Petroglyphs indicate that the Sahara was a lush region before it became drowned in sand, perhaps by a swarm of comet fragments that became electrically fractured into sand while strafing the ionosphere, or even the atmosphere, in an event that no one lived to tell about.

Northern boundary of the ice age safe zone

Relative solar irradiation

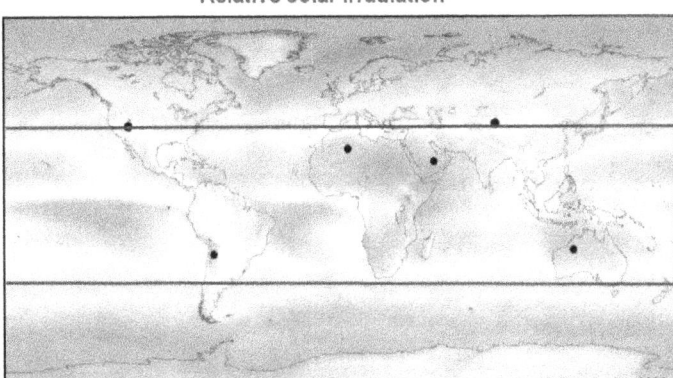

Since the Sahara was lush before this time and supported a civilization during the times of the inactive Sun, the northern boundary of the Sahara appears to be also the northern boundary of the ice age safe zone, coinciding with the thirty-five degree latitude.

*The safe line

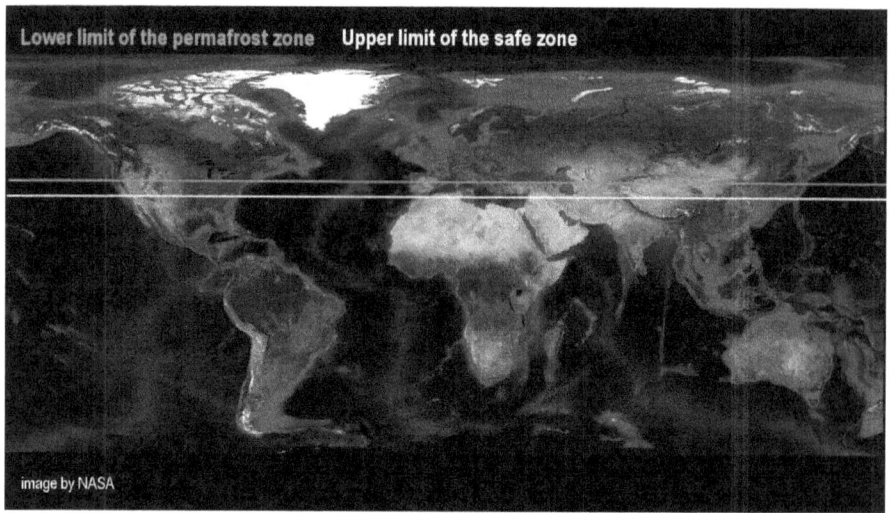

Lower limit of the permafrost zone   Upper limit of the safe zone

image by NASA

The safe line stretches across California in the USA, north of Los Angeles, and from there through the middle of the Mediterranean Sea, and in the East it cuts through the middle of China near Lianyungang, and through Japan south of Tokyo. Anything north of that, typically along the 40 degree line, from Beijing to Madrid to Philadelphia, was permafrost country during the last ice age.

This means that large portions of the Earth become 'difficult,' if not impossible, to live in during the solar inactive times. This includes primarily Russia, China, Canada, the USA, and Europe, which share a common problem, and thereby are natural partners for building the required solution.

56

A perfect zone free of hurricanes

When agriculture becomes disabled beyond the 40 degree latitudes

The World Development Project

Creating far-flung floating agriculture modules across the equator
between the hurricane zones, connected to a network of intercontinental floating bridges

Wikipedia / NOAA

If the solution requires to place agriculture onto the sea, since land
is scarce in the tropical areas, a perfect zone exists for doing this
right along the equator. A narrow zone exists along the equator that
is free of hurricanes. Placing agriculture onto floating modules
made out of basalt and produced in nuclear powered, automated
industrial processes, is probably more easily accomplished than
greening the Sahara in laborious manual operations. The same is
evidently also true for the automated production of complete
housing-modules, as a necessary and free infrastructure to mobilize
our humanity.
Some protest here, that this is too hard to do.

Floating agriculture is small stuff

If the people of a small culture, as far back as 12,800 years ago, were able to mobilize the economic resources to cut 4 million stone blocks out of the bedrock, weighing several tonnes each, transport them to the building site, and to place them with precision on a structure with a steep slope reaching 480 feet into the sky, in comparison with that, the building of bridges across the oceans with floating agriculture and floating pre-manufactured cities, is small stuff, considering the power of high-temperature automated industrial processes, motivated with nuclear power. A single 1 gigawatt plant should be able to produce 2,000 housing units an hour. We can have a whole new world coming online on this basis with comparatively little effort. No one needs to starve or perish when the Sun becomes inactive.

That's what the pyramids are telling us.

The future of Canada, Russia, and Europe

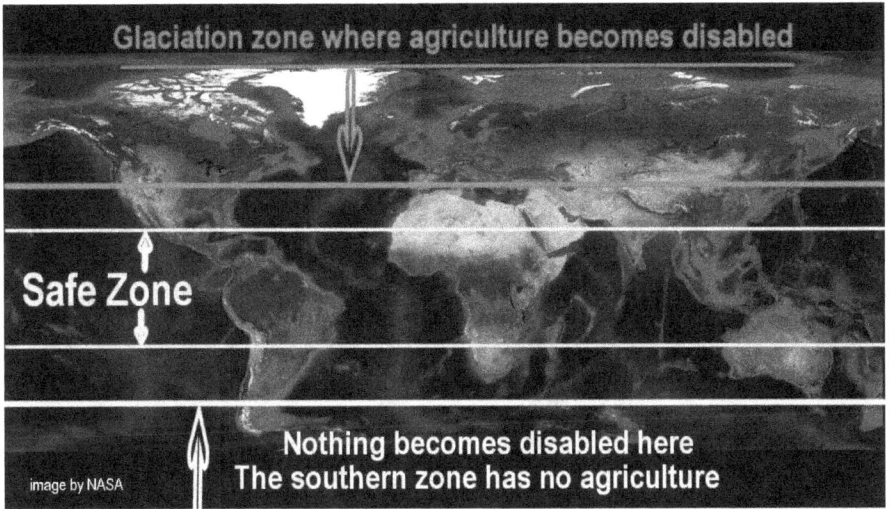

This means that the future of Canada, Russia, and Europe is logically located on the sea along the equator, and to a lesser degree the future of China and the USA. It also means that we better get busy soon, to transform our world.

Summer Solstice Sunrise over Stonehenge 2005

wikipedia

Another large construction project from ancient times tells us a similar story. The Stone Henge project, like the pyramid building project, reflects features that the ancient builders must have seen in the sky, as the features that were constructed, replicate critical aspects of high-power plasma physics that were only recently discovered in laboratory experiments.

A ring of 56 chalk pits

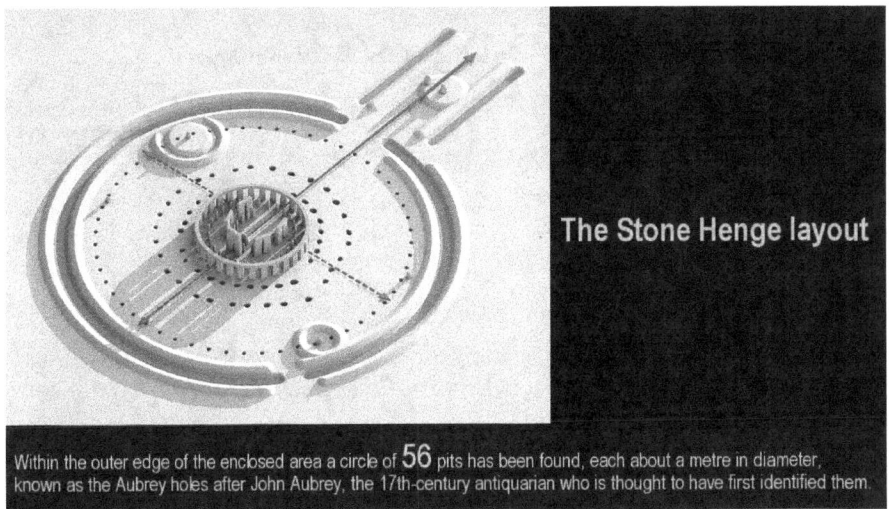

The Stone Henge layout

Within the outer edge of the enclosed area a circle of 56 pits has been found, each about a metre in diameter, known as the Aubrey holes after John Aubrey, the 17th-century antiquarian who is thought to have first identified them.

The Stone Henge monument features a ring of 56 chalk pits, a meter wide and three quarters of a meter deep, name the Aubrey holes in honour of the discoverer of them. Their purpose remains a puzzle unless one connects them with the sky.

A lab experiment, published in 2003

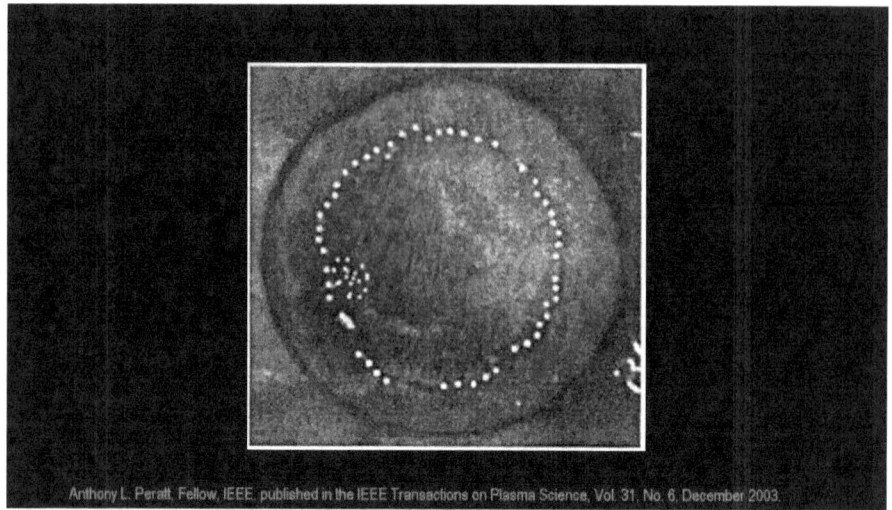

Anthony L. Peratt, Fellow, IEEE, published in the IEEE Transactions on Plasma Science, Vol. 31, No. 6, December 2003.

It was discovered in a lab experiment, published in 2003, that "A solid beam of charged particles tends to form hollow cylinders that may then filament into individual currents. When observed from below, the pattern consists of circles, circular rings of bright spots, and intense electrical discharge streamers connecting the inner structure to the outer structure." The maximum number of the filaments has been found to be 56.

A ring of 56 distinct plasma filaments

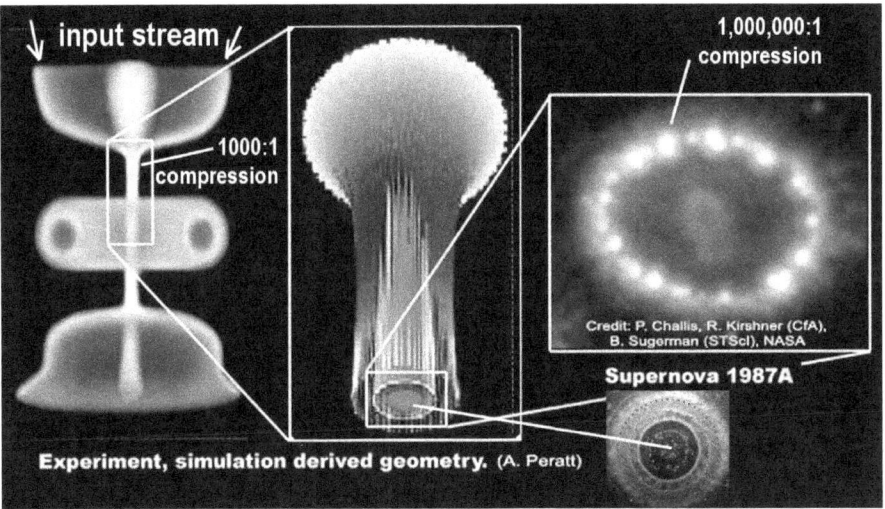

Another experiment shows that the concentrated plasma that flows between the two complementary bowl structures of the Primer Fields, forms a ring of 56 distinct plasma filaments, which in the real world, under extreme conditions, appear to have been visible in the sky.

It has been noted that over long distances the filaments merge in groups of two or three, as can be seen in this image of the Supernova 1987A.

The image shown here is giving us a cross-section view of a strong Birkeland current flowing in space.

A very-high power plasma stream

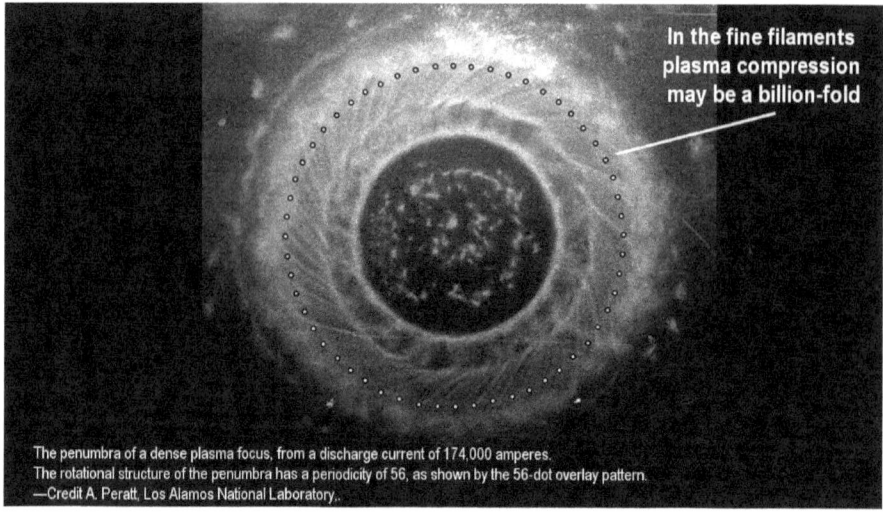

In the fine filaments plasma compression may be a billion-fold

The penumbra of a dense plasma focus, from a discharge current of 174,000 amperes.
The rotational structure of the penumbra has a periodicity of 56, as shown by the 56-dot overlay pattern.
—Credit A. Peratt, Los Alamos National Laboratory,.

In another experiment a more perfect cross section image of a very-high power plasma stream has been recorded that shows the structure of the 56 plasma filaments self-aligned into a circle, and streamers flowing into two other circles, and so on.

It is surprising to note

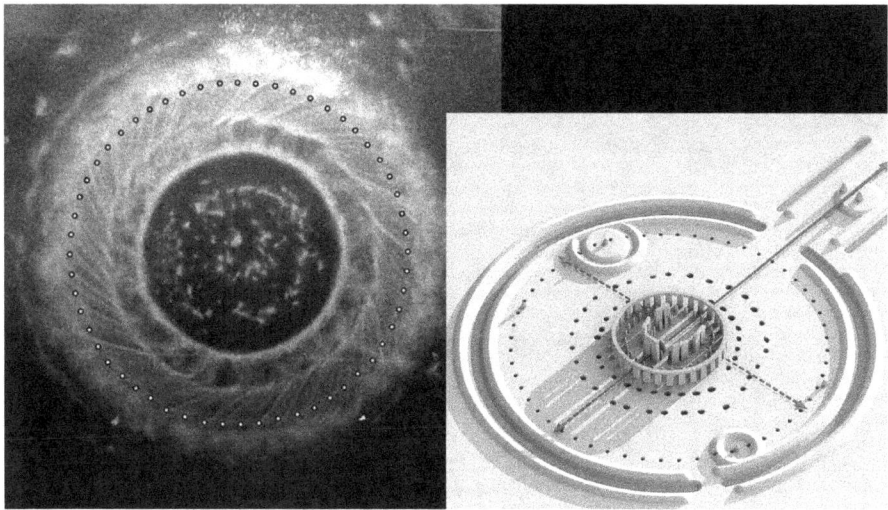

It is surprising to note how closely the layout of the Stone Henge monument, built so long ago, matches the plasma flow patterns that became visible only recently in experiments produced in the laboratory.

The surprising similarity that we find in these two cases suggests that the plasma-flow patterns had been seen in the sky in ancient times, probably seven or eight thousand years ago in times when the interglacial Sun was at it peak power level. The pattern that was seen then, which was replicated in the monument, would have been the outflow pattern of the concentrated plasma that was at the time visibly focused onto the Sun.

We no longer see these patterns

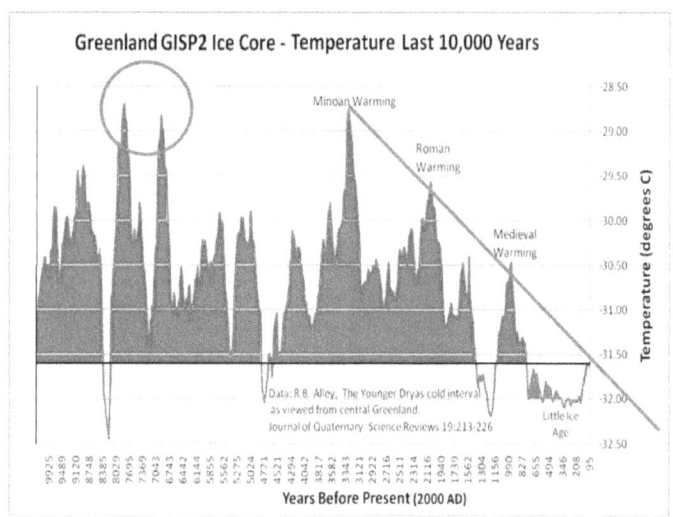

That we no longer see these types of patterns, is the natural consequence of the weak plasma environment today. There is simply not enough density left remaining for any of the plasma features forged by the Primer Fields to be visible. Soon, the Primer Fields themselves will collapse and the Sun become inactive again for long intervals, as it had been during the previous ice age period.

Politics versus science

** The title should be: Science Uplifts Politics, But until we get there, the title remains, 'Politics versus Science.'

Humanity urgently requires the ice age challenge

Annihilation is assured

500,000 times Hiroshima in one hour

Castle Bravo - the first U.S. test of a dry fuel thermonuclear hydrogen bomb - March 1, 1954 at Bikini Atoll, Marshall Islands

It may well be that humanity urgently requires the ice age challenge that now stands before us, to awake the world from its current rut before the 'sleep of reason' causes humanity to become extinct by the nuclear war that the world's leaders of empire have been building weapons for, for more than 50 years already. It would only take an hour and a half of thermonuclear war to create the conditions that causes the extinction of humanity and most forms of life with it.

A spiritual task, more than a political task

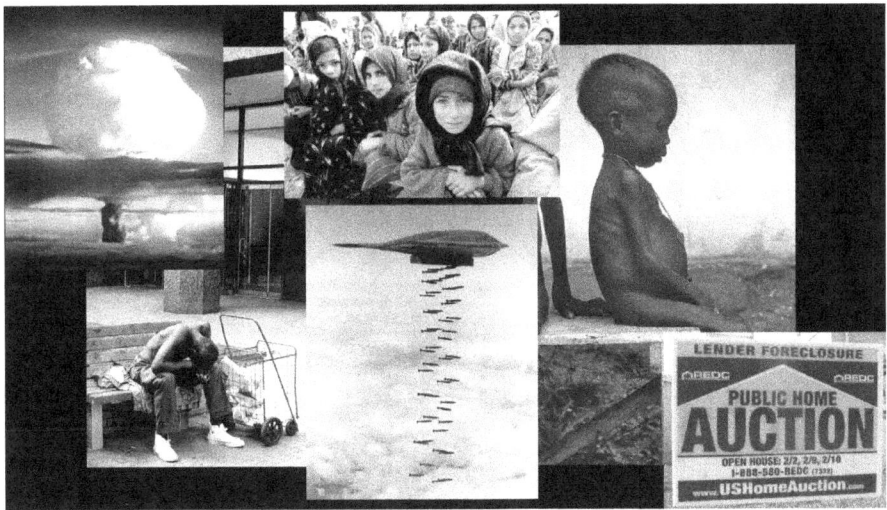

We need to raise the value of our humanity sufficiently high to take the steps required to cleanse the human landscape of the terror of nuclear war and everything that is linked to it. This is a spiritual task, more than a political task. The correlation of cosmic-ray flux with solar activity, indicates that the advance of civilization is a spiritual task. This alone should inspire us to pursue spiritual development as the highest priority objective of modern time, in order to become sufficiently spiritual minded to protect ourselves.

We wield weapons out of weakness

We wield weapons not out of strength. We wield weapons out of weakness. We wield them because of our inability to face one another as human beings. Let's hope that the ice age challenge will awake us enough to consider what human living is all about, before we throw it away without ever having really lived.

We face a choice

We face a choice therefore. The choice is between the system of empire for which all wars are fought, and the platform for freedom on which human culture and human development depends. So far humanity stands impotent and tied to the ground. It stands latched to the system of empire where it hails the weapon and throws away its humanity, as if it was already dead, unable the choose life and to choose it more abundantly.

Humanity still lives in the Roman age

Universal terrorism!
The 'Religion' of the Empire of Rome

The Christian Martyrs' Last Prayer by Jean-Léon Gérôme (1824–1904) - Roman Empire, Wikipedia

In many respects humanity still lives in the Roman age, reflecting the grand fascist arena of monetarism married to terror, inhumanity, and depopulation. Large segments of humanity are presently being killed under the new Roman banner, and brutally starved, torn, and despised, while much of the world looks on silently, betting on the indexes for profits.

We simply starve them to death

While we no longer burn people alive by the dozen as in the grand arenas long ago, we simply starve them to death at a rate of 300,000 a day by burning their food in automobiles in the form of biofuels.

*The ice age challenge as a new paradigm

We need the ice age challenge as a new paradigm for living, in order that we may heal us of the tragedies that we have allowed to come upon us.

If we fail to become human above everything else; above politics, economics, religion, status, and sex, and fail to see ourselves as a single humanity honouring one another in love, we find ourselves unfit to meet the ice age challenge, which, technologically is not a great challenge to meet.

Only we lose by our failing

Antarctica - the coldest place on Earth
Sentinel of the Pleistocene
The face of the world of 'normal.'
Photo by Vincent van Zeijst / wikipedia

However, if we fail the ice age challenge, as we have failed in the past by not preventing the great wars, destruction, and looting carried out in the name of empires, then the coming ice age will bury us, with few exceptions. Perhaps then, in a thousand generations, or a million years, a new humanity will rise again and will not fail those challenges that we presently have no intention to even address. The universe won't be cheated by us failing ourselves. Only we lose by our failing. But why should we fail?

To rekindle that flame in the heart

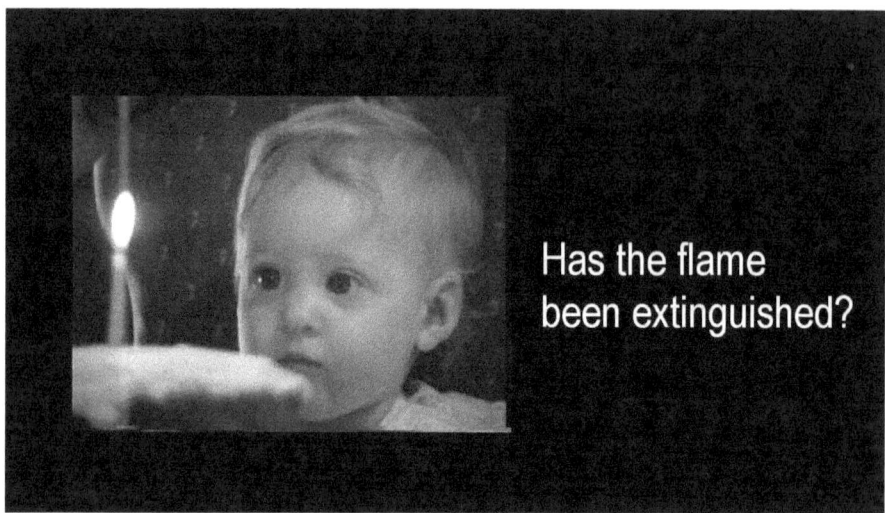

Has the flame been extinguished?

We only need to rekindle that flame in the heart that rekindles our humanity and our human productive and creative power, and the beauty of our human soul and its scientific honesty. The world has become a desert landscape in all of these areas. We need to begin by turning the desert into an oasis across the world. The fading little flame of today, needs to be nourished to become a great fire in civilization. Surely, we can do this.

Not even the sky will be a limit

(C) Corel Corp

When we get to this point, not even the sky will be a limit for us, much less the conditions that we find on the Earth. But will we do this?

Will we dare to step up onto the wings and fly?

Will we dare to step up onto the wings and fly? Will we trust ourselves, for the great task? These remain still-open questions. Evidently the greatest danger before us, is not the astrophysical one. This one we can deal with fairly easily and protect ourselves from the astrophysical changes.

*The greatest danger that we face

The greatest danger that we face, then, is that we will do nothing, that we won't respond to the astrophysical challenge before us and die of starvation by not having created the infrastructures for our food supply that enables us to live richly in the dimmer world.
The danger is that we keep up the present course, that we continue debating global warming and manmade climate change, and that we continue to bow to the monetarism of empire till the Sun turns off and goes into its sleep mode.

Brainwashed by the choruses of the professional scoundrels

We have been brainwashed to commit ourselves to the present tragic course by the choruses of the professional scoundrels that the masters of empire have bought with their money bags to keep the dream of empire alive.

People who have sold their soul for a song

Those, who sing the empire song, are people who have sold their soul for a song and have turned civilization into a desert of starvation already, before the big challenge even begins.
That's what the global warming hoopla; the nuclear war terror; and depopulation are all about, standing against the future of humanity, aren't they?

Brainwashing to keep the internal-fusion sun theory alive

A similar type a brainwashing has also been applied to keep the internal-fusion sun theory alive at all cost, as it serves the policies of empire well. It prevents economic development, energy development, and scientific development. Humanity is being kept tied into knots by debating the internal-fusion sun theory, because the outcome of the brainwashing promises to facilitate the cherished objectives of empire, which is to eliminate 6 billion people from the face of the planet from the present 7 billion to about 1 billion. That's the officially stated policy for depopulation. Would you like to be 'depopulated?'

Children to become depopulated?

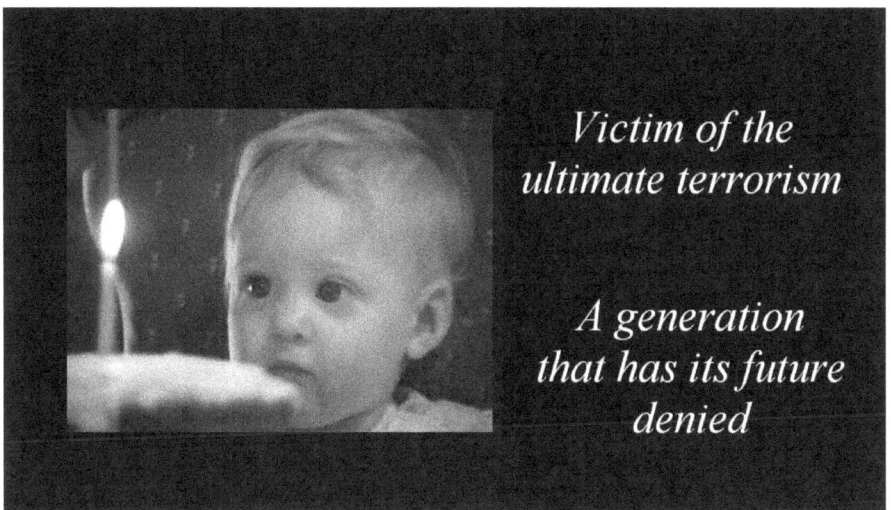

Would you like to see your children to become depopulated? This is what will happen if the ice age challenge is denied. If your answer is, yes, to the depopulation question, then sit back and do nothing, as you may do together with most of humanity, and you will have what you wish for. There is today little evidence on the horizon for a breakout from society's commitment to its small-minded thinking that has been imposed on it, such as in the context of the global warming scam and the internal-fusion sun.

The Sun is an electrically powered star

The train of the internal-fusion sun continues on track, even while the evidence is monumental that the Sun is an electrically powered star.

Behind the scene of denial

Electric arc by a switch failure on a 500 Kv transmission line

Solar prominences

Eldorado Substation near Boulder City, Nevada
www.komar.org/christmas/faq/electrical_overload.html

NASA

However, behind the scene of denial of the evidence that we see - a denial that is political and artificial in nature - stands the humanity that we are - a highly intelligent species that will always remain what it fundamentally is, which tends to assert itself in times of deep crisis.

The intelligence of humanity

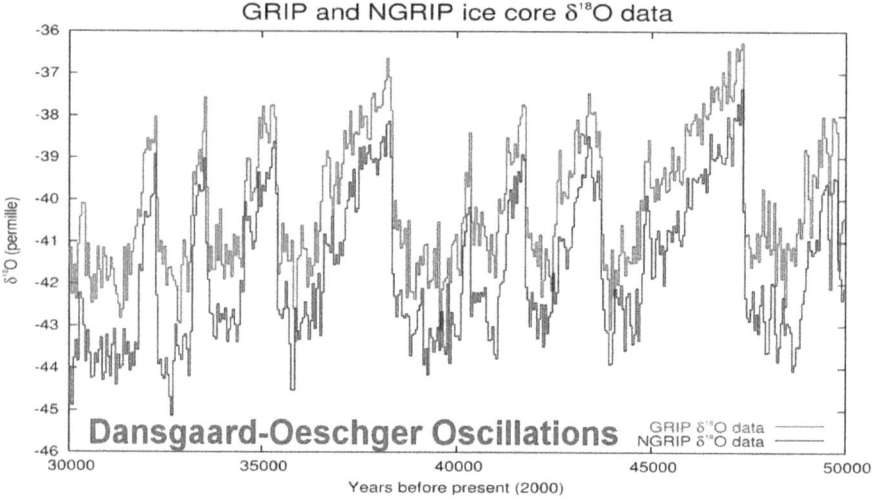

The intelligence of humanity will make its claim, slowly as this may unfold, and stand for truth, and will thereby prevent the greatest catastrophe of all times from occurring, which would result with certainty if the astrophysical processes that evidently loom before us, would remain too long ignored and not be responded to.

Break through the fog of political games

CERN - CLOUD project - Jasper Kirkby

It is inherent in the design of humanity that the leading-edge thinkers will break through the fog of political games and open their eyes to the mountains of bodies of evidence that astrophysical processes have a powerful impact on earth.

Alert scientists and truth-seekers

It is inherent in the intelligence that we embody and reflect as humanity, that the most alert scientists and truth-seekers do see the phase changes before they happen, by understanding the processes involved, and that they will inspire society to move with the requirements of the future that can be shaped with intelligent actions.

Scientific progress the hallmark of humanity

Scientific progress remains the hallmark of humanity. Even if the options are at times obscured, and the paths seem tied into knots, enormous scientific efforts have been made, and continue to be made, to push forward our understanding of the universe and its operation, and of ourselves in the unfolding process.

Unlimited energy resources

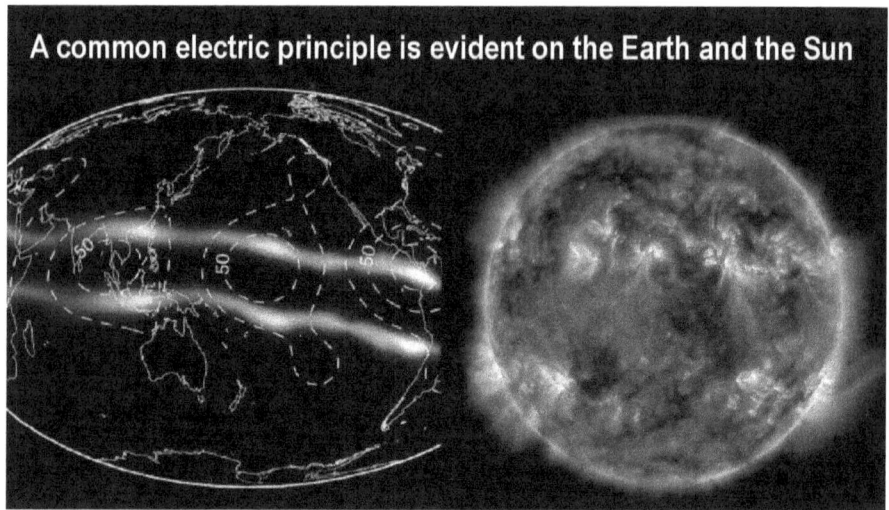

A common electric principle is evident on the Earth and the Sun

The more we move forward with our exploration on this front, the more will a great renaissance take shape before us, with unlimited energy resources that give us access to unlimited building materials and food supplies, all with the kind of density that enables us to respond to the changing solar system, which then can no longer endanger our civilization nor hinder its advances.

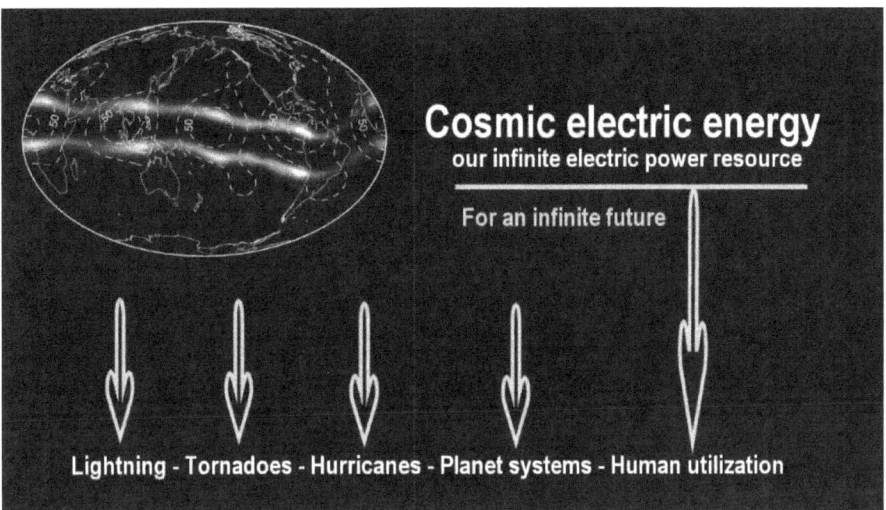

In fact these potentials stand before us right now, to be realized now, and be developed fully before the Sun reverts to its inactive state as it did during the previous ice ages.

In the previous ice ages the great potential that we have today to create a new world, did not exist, but it does exist now. As an intelligent species, the most advanced that ever existed on the Earth, we will not squander this potential.

Maybe this is what the Universe recognized and put to our credit

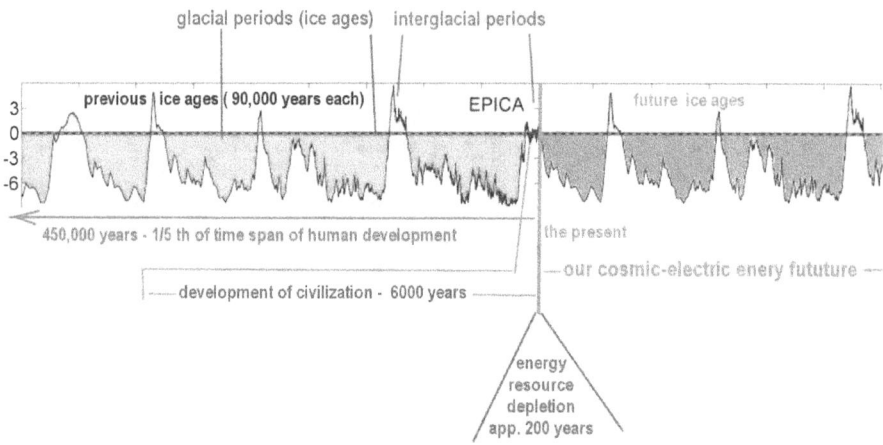

Maybe this is what the Universe recognized and put to our credit. Humanity had proven its worth before the Little Ice Age began.

To create a society without empire

During the Little Ice Age, for the first time ever, a process was set in motion to create a society without empire on the distant shores of America.

The city on a hill

Freedom to develop is our future

The pilgrims' determination for the first time in history to raise the finger to empire eventually resulted in the founding of the USA as the first sovereign nation state on the planet - the city on a hill that the eyes of the world were indeed fixed upon.

The remarkable achievement that had begun, which was a radical new development in civilization - for a people to collectively take a stand against the system of empire - would likely have been lost together with much of humanity if the Little Ice Age had progressed to become the big Ice Age. But this train to near extinction was blocked with a major Dansgaard-Oeschger pulse that ended the Little Ice Age and extended the interglacial period by a few more hundred years from the 1700s on.

The greatest economic and scientific development

Ulysses was launched from the NASA Space Shuttle Discovery on Oct. 6, 1990

NASA

The extension of the interglacial was extremely critical for the development of civilization. It set the stage for the greatest economic and scientific development the world has ever seen, even though much was squandered away in recent years.

Warm climates after the Little Ice Age

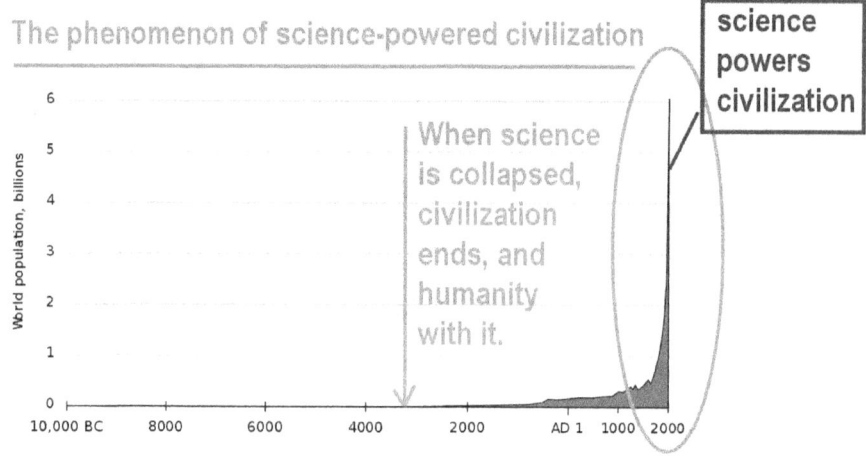

The 200-year period of warm climates after the Little Ice Age, which was by its timing evidently a major Dansgaard-Oeschger pulse, gave us the population increase that is required as an economic base for carrying out the redevelopment of the world in preparation for the coming ice age period when the Sun goes on and off for typically 90,000 years. Was this ideal coincidence, accidental? Was it a gift from heaven, to save the world? Or was it earned by humanity?

The war against empire has not yet been won

While the war against empire has not yet been won, the challenge of having to prepare the world for the environment of the dimming Sun will bring back to light the great achievements of the past and thereby swing the balance from empire and its wars and its money, to freedom and to the fulfillment of the common aims of all mankind.

This trend away from empire towards the freedom of humanity started in the 1620s with the Pilgrims that settled in Massachusetts, who came there to set up a New World on the American shores, far from empire. The Peace of Westphalia that occurred later in 1648, to end the Thirty Years War, also started a movement in the same direction in Europe, which still stands to some degree as the foundation for modern civilization.

Society's self-directed spiritual development

On this same type of path, as the power of society's self-directed spiritual development is stepped up far above the default train, the threat of nuclear war will vanish and become forgotten history as though it never existed, together with all the ugly garments that empire still parades: its monetarism, terrorism, fascism, greed, slavery, genocide, poverty, hatred, lies, and so on and on.

To meet the human need

Wheat harvest on the Palouse, Idaho, USA

Food for eating!

wikipedia

The requirement to meet the human need in the grandest style possible and with the greatest freedoms unfolding from it, has been the basic dream and hope in every religion that ever was, which is a dream that has so far never been fully realized, but might be realized at last as we step forward in our self-directed progression from religion to the scientific realization of the grand humanity that is inherent in us all, in which we are One.

*Challenge of the dimming Sun

The astrophysical challenge of the dimming Sun puts a new type of challenge before us that inspires new paradigms, since the old paradigms no longer apply.

Something to celebrate

Philharmonic Orchestra of Jalisco
(Guadalajara, Jalisco, Mexico)

wikipedia

 This is something to celebrate. It inspires us to reclaim our humanity, raise it up anew, explore it in new novels, new poetry, and new songs, on the road of our intentional development of our unlimited potentials.

This makes our future bright in spite of the dimming Sun, or because of it, for the challenge it imposes.

Harvest is Seedtime

Harvest is Seedtime

Seeds, wind-blown
Carriers of a secret still unknown
Poems in the words of nature
Sentinels of an Intelligence yet unseen
Prophets of the enduring
Apostles in an endless landscape

A poem from the novel
Endless Horizons
by Rolf A. F. Witzsche
of the series of novels
The Lodging for the Rose

Harvest is Seedtime
Seeds, wind-blown
Carriers of a secret still unknown
Poems in the words of nature
Sentinels of an Intelligence yet unseen
Prophets of the enduring
Apostles in an endless landscape

Discovering Love

Discovering Love

Seeds, wind-blown

Poems in the words of nature

Sentinels of an Intelligence yet unseen

The melody of nature - what a song!

The melody of nature - what a song!
Whispers of a splendor grander than the heavens
Like seeds are we - we whisper too
Seeds bearing gifts for the world
Gifts wrapped up in sunshine
Gems are we - unfolding a majestic song!

The melody of nature - what a song!

Whispers of a splendor grander than the heavens

Like seeds are we - we whisper too

Seeds bearing gifts for the world

Gifts wrapped up in sunshine

Gems are we - unfolding a majestic song!

Lu Mountain

Lu Mountain

Whispers of a splendor grander than the heavens

Like seeds are we - we whisper too

Seeds bearing gifts for the world

Listen to the song

Listen to the song
Listen to the heart
Listen to the silence where strands of love unfold
Listen to the symphony of our humanity
In this symphony we are One
One with the Universe itself.

Listen to the song
Listen to the heart
Listen to the silence where strands of love unfold
Listen to the symphony of our humanity
In this symphony we are One
One with the Universe itself.

Flight Without Limits

Flight Without Limits

Listen to the song

Listen to the heart

Listen to the silence where strands of love unfold

Brighter than the Sun

Brighter than the Sun

Listen to the symphony of our humanity

In this symphony we are One

One with the Universe itself.

www.ingramcontent.com/pod-product-compliance
Lightning Source LLC
Chambersburg PA
CBHW071822200526
45169CB00018B/600